I'm a newcomer. But it seems to me that Lizzie-Lou was born in a saddle. Why, she and Lightfoot are tiptop cattle herding partners. Day in and day out, they work the range together. Sunup to sundown, you won't find one without the other. They're like grits and gravy, that cowgirl and her cow pony.

"A cowgal's best friend is her pony pal," Lizzie-Lou says.

And Lightfoot is no kind of simple, commonplace pony, anyhow. Likewise, Lizzie-Lou is no plain, everyday cowgirl. I may be a newcomer, but I know what I see myself.

Lightfoot is the fastest pony that ever was! He's faster than lightning chasing the wind. Why, Lightfoot runs so fast, Lizzie-Lou's own shadow can't keep up!

And doesn't Lizzie-Lou have her own talents too! She's the loopy lasso expert of these parts. If it's on the loose, she can loop it with her lasso. Some say she could rope a rainbow out of the sky. But it wasn't always that way, the old-timers will tell you.

"Once upon a time, talk like this would be poppycock," they say.

Now, I'm a newcomer, so I don't know for sure. But I do know, sure as Lightfoot is fast, Lizzie-Lou can lasso anything. I've seen both firsthand.

But old-timers say: "Lightfoot once was a slowpoke." They also say: "Time was when Lizzie-Lou couldn't lasso a stick if it was standing still."

Well what caused the turnabout? Here is what an old-timer told me:

"It was their first roundup together. Lizzie-Lou was a tenderfoot, and she couldn't rope a thing. Lightfoot was tiptoe slow. So roundup lasted late ... with no cow caught.

Everybody (cattle, cow pony, and cowgirl) tuckered on out. But Lizzie-Lou and Lightfoot were truehearted and wouldn't quit. No! So pretty soon everyone everywhere was sleepwalking.

Fact is, one cow went right off into the sky, bumped into the moon, and knocked it out of place.

Moon fell down and hit Lightfoot, top side of his head. Moonstruck, Lightfoot fell into a tailspin. Well, Lizzie-Lou—tossed roundabout on top—looped everything along the way. She put the moon back up, got the lost cow down, and roped a few stars too.

So you see, Lizzie-Lou and Lightfoot skyrocketed to fame in their star-studded roundup. By daybreak those two were giddy—and mighty proud of what they had done! With starlight in their eyes, they have never slowed down since … that cowgirl and her pony."

That's the tale the old-timer told! … Sure sounds true to me.